IGNATIUS BEARD

Mind-Blowing Space Facts

Amazing Details you never knew about our Incredible Universe

Look deep into nature, and then you will
understand everything better.

ALBERT EINSTEIN

Contents

1

Introduction

Welcome to the mind-blowing space facts! I'm so thrilled you found this book, in which I hope you gain greater insight on a wide array of Cosmology topics, like what most likely happened right after the Big Bang, what strange phenomena is occurring right in our Galaxy, and what's the fate of our Universe? As widely encompassing and complicated as these topics sound, I also hope to present each one in an easy to read format, yet equipping you with so many facts that you can impress your friends!

Before we get into these revelations, I'd like to introduce myself. My name is Ignatius Beard, also known as Iggy to my friends, and I possess a deep love of all things space and NASA. This love propelled me to pursue and completed my undergrad with a BS in Physics in 2012, from the wonderful University of North Carolina at Chapel Hill (Go Tar Heels!), and furthered my passion for Math with a MS in Statistics in 2017, from the University of Southern Maine (Go Huskies!).

As you may already know, there are perhaps infinitely more facts about outer space, countless more than this book will cover, but I'm still excited to at least share a few of my favorite facts with good, but not too much, details regarding each. These topics presented will be those that captured my mind in University, and specifically one fact, that I won't spoil yet, blew my mind so much in high school, that it became my topic for Junior and Senior projects (you'll know which one it is)! My hope in this work is that you'll gain a greater perspective on how incredible our Universe actually is, and I'm sure you'll learn at least one detail that will blow you away.

As mentioned before, the content of this book will serve as a brief introduction into these topics and concepts, since each could very well be expanded into their own books. My goal is to present fascinating details that the everyday person may not readily know. I'll attempt to avoid intense jargon also, so don't worry, as I want this book to be so understandable that my mother, who's not scientifically inclined at all, can understand and fully appreciate what's inside.

Now that you know a little more about me, let's explore the Universe together!

2

Everything you see came from Stars

When deciding which fact to share first, I couldn't help but start here! As shocking and unbelievable as this may sound, everything you've ever seen, heard, smelled, touched, even tasted, and I mean everything, has come from an exploded star! This includes everything we see on our planet Earth, from the fine sand in the Sahara Desert, to every water molecule in the Gulf of Mexico, from the sandwich in your refrigerator, to the carbon in the oak tree in your backyard. If you're asking yourself, "How in the Universe is this possible?" I'll say you're halfway there. This truth comes from our understanding of the early Universe, in which Star tens of times the size of our Sun lived short lives, burning through their fuel so fast, and exploding heavy elements out into space!.

It's truly fascinating when you take a step back to think. Other than Hydrogen and helium, the vast majority of other elements have stars to thank for producing them. You may be asking yourself, "Well what about comets? Those cold, icy/rocky projectiles zooming through our solar system couldn't possibly come from stars, right?" Even these ice balls have to credit stars, as they emerged from the leftover debris in our solar system, this debris that, you guessed it, came from birthing our very own Sun. The origin of all this heavy matter traces its life to these stars that existed early in our Universe, consuming their energy so quickly that they destabilized to the point of going supernova, launching iron, cobalt, and other elements that are so useful to us today.

So now, every time you hold your favorite gadget in your hand, hike through luscious and vibrant nature, or experience your favorite form of entertainment, take a moment to appreciate these ancient stars!

3

Everything you see is in the Past

Now, this truth never ceases to fascinate me, even after processing the complex math that explains why. When you go outside on a clear night and see your favorite star, what you're actually viewing is how that star, or galaxy, or other bright object, appeared a measurable amount of time in the past.

This isn't just exclusive to viewing far out galaxies either. Taking our local solar system for instance. If we go outside midday and look up (very carefully) to our Sun, what you're seeing is what our star looked like 8 minutes ago! What's wild about this is that we wouldn't be able to detect any change to the Sun for 8 minutes. Let's say that our local star just vanished out of nowhere, or exploded randomly. We literally wouldn't know or couldn't tell for 8 minutes!

As we view objects further out in the Universe, we peer more into the past. Our closest Neighboring Star outside our Solar System, Proxima Centauri, is 4.2 light years away, and when we view this star, we're seeing how this star appeared a little under 4.2 years ago, when the light was emitted. This means that the star could have disappeared 2 years ago, and we'd still be viewing it in the sky for 2 more years!

What's even more incredible, is that as of writing this moment, scientists were able to see what they believe is the farthest object ever from our planet, a potential galaxy dubbed 'HD1'. This object has a distance of a little over 13.3 light years away from Earth, however due to Universal Expansion, HD1 can very well be roughly 33 billion light years away from Earth!

So, why do you think that is? What causes us to see these outer space objects in the past? If you're puzzled, don't worry, an upcoming fact will calm your worries!

4

There's Surprising Evidence for the Big Bang

Many of you, if not all, who have picked up this book, have at least heard of the famous "Big Bang Theory" (No, I'm not talking about the show, though its humor did help me get through my Grad School!). However, some of the details surrounding the Big Bang, specifically right after that Bang, may not be well known, but yet are truly incredible. Imagine an environment so hot that even atoms couldn't exist, or density levels greater than the supermassive black holes! This is what's witnessed in our Universe's beginning and what breaks down our Laws of Physics.

What I find exciting is that we live in a time where our measurements and tools are constantly improving. With state of the art telescopes in use, such as the famous James Webb, our efforts in Cosmology are allowing us to see further into the past and uncover brilliant details of the early Universe.

In fact, you know the galaxy HD1 that I mentioned earlier? Well this galaxy was in fact discovered by the James Webb Telescope, the oldest object on record. The presence of this object is placed just 330 million years after the Big Bang! Who knows what else we'll find in the next 5 years?

5

The Laws of Physics Evolved after The Big Bang

When I worked through intensely difficult quantum, particle, and Newtonian physics problems and projects in my undergrad, when I staying up constantly, asking professors for help, wrapping my head around Hamiltonian, Fourier, and Laplace transforms, even praying a bit in between exams, there were moments where I wished that Laws of Physics didn't exist. Possibly for the first minuscule fraction of a second after the Big Bang, between 0 and 0.0000000000000000000000000000000000000000001 , or 10^{-43} seconds, I would have had my wish. Even though I would have been an energetic wave function of goo, at least I wouldn't have exams!

This unbelievable fractional span of time in our Universe is known as the **Planck Era**, and surprisingly, the Laws of physics as we know them today cannot explain what took place in the Universe during this time. The best hope for insight would be a "Theory of Everything" to help fill in the gaps in our knowledge, connecting the behavior of all matter and energy, under all conditions, big and small, together. This theory would instantly be one the greatest discoveries ever, and is currently the aim of many scientists. Even though it's nearly impossible to picture this time period through our Physics lens, it's fascinating to think that there are still mysteries to discover.

To really appreciate this mysterious origin, I find it helpful to briefly introduce these what are known as the 4 fundamental forces. Don't worry I haven't forgotten my promise to keep it all understandable. These forces essentially shape the Universe that we see today.

Gravitation Force:

The first Law we visit is the Gravitational Force, which I'm sure is one we could have guessed, that governs how two objects attract each other. As simple as this one sounds, the force of gravity gets much more interesting for objects at a larger, more celestial size, as Einstein's General Theory of Relativity comes into play. At this scale objects through its mass create bends in actual space, expressed more properly as space-time as this bend also affects time! All objects affect space-time, including you and me, but only at a negligible scale. Immensely large objects, however, like supermassive black holes, really showcase this force and bending. I'll move on shortly, but remember this as I show the other forces: Gravity is actually **the weakest** of the fundamental forces!

Weak Nuclear Force:

Now you may be wondering "But I thought Gravity was the weakest force?" Believe me I know it could appear counter intuitive, and this is how Scientists like to name things sometimes, but in an attempt to clear the naming up, Gravity is the weakest force. The **Weak Nuclear Force** is next in line on the power scale. This force describes the behavior of subatomic particles, known as quarks and leptons. It operates on the shortest range of all the fundamental forces, and is essentially what's responsible for radioactive decay in elements, and quark behavior in particles which can change a proton to a neutron and vice versa.

Electromagnetic Force:

Considered the Second strongest force, the Electromagnetic (EM) force is another one you've heard of before. This force describes the behavior of electric and magnetic fields, which powers much of our modern life, from movies, to gadgets, to our energy sources. It might be surprising to you to think of this force as much much stronger than Gravity, but I recall a quick demo in my physics class in high school, where my teacher, Mr. Patterson, took a magnet and pulled a paperclip off a desk. While this seems mundane and expected, that 1 inch magnet simply overcame the entire gravitational pull of the Earth on that paperclip! While this EM force is very strong, its effect is neutralized through balanced positive and negative charges.

Strong Nuclear Force:

Now for the strongest force in the Universe, the Strong Nuclear Force! This force keeps protons and neutrons together in an atom. It also works on a deeper level, binding the quarks that make up protons and neutrons together, through the subatomic gluon particle (just think of"glue"). This is notably 100 times stronger than the electromagnetic force, but only has a range of the atomic level.

Now that you're equipped with an overview of these forces, let's revisit the fascinating beginning of our Universe.

Universe Timeline (0 seconds after Big Bang):

0 to 10^-43 seconds: The **Planck Era:** Nature of forces Unknown

10^-43 to 10^-35 seconds: **Gravity** Becomes its own force. The 3 remaining forces remained combined as part of the **Grand Unified Theory**

10^-35 to 10^-32 seconds: **Strong force** Becomes its own force: The electromagnetic and weak force are still combined as the **Electroweak force**

10^-32 to 10^-10 seconds: **Weak Force** Separates, now all forces are separate.

After **10^-10 seconds**, before even a single second passed in the Universe, the fundamental forces that we know today are present. That mysterious **Planck Era** causes us to wonder however what strange forces were present during this time, not even taking into account happened 1 second **before** the Bang!

6

The Universal Speed Limit

Have you ever wondered how fast we could ever travel with technology? Our world is constantly making things faster, from cars to our internet to delivery. We've had airplanes that break the sound barrier for over half a century now, so could we one day have a spacecraft that zips from one end of the Universe to another? Maybe we'll have spaceships moving in terms of parsecs per hour!

Well, I learned the answer in my modern physics class, and while I may not have the answer you're looking for, I do have another fact: Our Universe has a natural speed limit for all material objects, which is the speed of light, measured at 300,000 Kilometers (or 186,000 miles) per second. In Physics this fact is also a constant denoted by the letter **c.**

This is the cosmic speed limit that nothing formed of matter can cross, and only light, or the massless photos that make up light, can reach this limit.

This also coincides with an earlier fact I mentioned before, as to why objects that we view in space are reflections of those objects in the past. The reason is that we rely on Electromagnetic waves, which include visible light, UV, inferred, and other frequencies, to view anything in our Universe, and since these waves move at the speed of light, we can only see the light from objects that reach us. I mentioned earlier that if the Sun instantly disappeared, we wouldn't know about it for about 8 minutes. That's because, even though sunlight moves incredibly fast, the distance to Earth is so far that the last ray of the sunlight, traveling 300,000 km/sec, would still take 8 minutes! Knowing that the light we see from objects in space travels to us at a set speed allows us to measure its distance away from us, and track how long ago in the past we viewed it.

That all being said, it does get interesting once we start approaching the speed of light. Say you borrowed a spaceship from a friendly alien, a spaceship which didn't have a special relativity shield, and without that shield, the ship didn't have a way to counter the special theory of relativity constraints for objects moving close to the speed of light.

Well, as soon as you cranked the boosters and accelerated the speed to a fraction of the speed of light, like say $0.40*c$ or $0.7*c$, your ship's observed mass would notably increase, forcing your spacecraft to supply more energy to accelerate as now it's pushing a more massive object. However this change in mass will eventually grow to infinity, meaning the energy required to push this object would also need to go to infinity, so that you and your craft can approach the speed of light. This phenomena is elegantly displayed is Einstein's most famous equation: $E = m*c^2$, where E is energy, m is an object's mass, and c is our speed of light constant. In it we see that the relationship of Mass and Energy is linked, with the speed of light always constant. While we on Earth don't feel this constraint first hand, it is present in the Cosmos.

Upcoming facts about the visible Universe that will Shock you.

At this point, you may be happy with yourself,since you now know a lot more about our Universe, congratulations! However, the next wave of facts may bend the mind a little more, challenging our natural assumptions. These next 2 topics really gripped me in high school, and helped propel my path to study Physics in College. So if you're ready for some truly mind blowing stuff, let's dive in!

7

Dark Energy: The Mysterious Substance that Fills the Universe

I remember first reading about this in my Junior year of High School, when we were picking research topics for our upcoming senior project. I knew I wanted something Space or NASA related, and while conducting a few google searches, I came across this crazy revelation: The Universe was not behaving at all as we thought it should. I started seeing blog posts about this **"Dark Energy"** that must be present everywhere to explain the unexpected being observed.

To fill you in, I first must provide background for this mysterious energy. Before the 1900's, we used to think that the Universe was in this equilibrium and static state, with its size being constant. All of this changed in the 1920's, when American Astronomer Edwin Hubble, through his Redshift research, discovered that galaxies were actually expanding away from each other. Not just in one direction either, this meant that all galaxies were actively moving away from each other. This finding was since accepted, and the popular scientific opinion was that, with our understanding of the forces of nature, and our ability to extrapolate, we could be confident that the current state of the Universe fell into of the 2 following scenarios:

1. The expansion of the Universe would eventually slow to a stop, and with enough gravitational force, began to shrink and collapse on itself, like a rubber band that's stretched past its equilibrium and recoils back.
2. The expansion slows down, and the gravitational force would be too weak to reverse it.

In both scenarios, expansion was expected to slow. This would make sense to us too now, as we know that gravity would act as a resistive force in this expansion.

What's almost poetic is that, in 1998, another groundbreaking discovery was made, this time by the famous Hubble space telescope, named after Edwin Hubble himself. Rather than the expansion slowing down, this discovery confirmed the opposite: our Universe is actually expanding **faster** than before! This expansion is so fast, that space is actually expanding faster than the speed of light! While this might sound like the Universe is breaking the crucial speed limit, but it's the space that is expanding at scale, not the matter in the space.

Expansion originally was a result so shocking that even Einstein attempted to correct his equation for the General Theory of Relativity by adding a Cosmological Constant to account for it. However, there was no reason found for this constant's existence in Physics, leading scientists to further wonder what's going on. This represents a result totally unexplainable with our current Laws and the matter we see in the Universe. Therefore there must be something else out there that we truly don't know. The fact is: The Universe is filled with an **unknown energy** that's accelerating its expansion.

While we can't explain or even see this energy, we can view the accelerated Universe, and when we factor in this expansion, we get some staggering insights. Creating a mass and energy model of the Universe, we find that all the visible mass that we see, and with our refined 4 fundamental forces that you've now learned, only accounts for 5% of this Universe!

To put it a different way, with our current instruments that peer far into the Cosmos, we can only see and understand roughly 5% of what makes up this Universe. When I first found these crazy numbers, I had to dig deeper. Without a doubt, this fact and the next one are the biggest and most worthy of the term 'Mind-Blowing!'

The breakdown is equally astonishing. Scientist were able to factor in the observable, normal matter into their calculations and factored in the mysterious phenomena, and if the Universe could be described as a pie, this **Dark Energy** would inhabit over **68%** of the Universe. This is staggering and only the best models right now show that this acceleration may have started ~7.5 billions years ago, with no concrete explanation.

There were attempts to explain this extra energy with a proposal that empty space actually contains energy, and as space expands into empty space, the energy present is not lost in the expansion. This proposal however didn't make it far, and instead other attempts to track down this energy have been employed.

So if **Dark Energy** takes up 68% of the Universe, and normal visible matter makes up 5% , where is the other 27%?

8

Dark Matter: The Universe is Almost entirely Invisible

Now comes the other side of the mystery coin, and the focus of my Senior project. The missing 27% of the Universe is credited to this invisible matter that doesn't emit any light, therefore practically invisible, yet is present in **every** galaxy, and primarily is responsible for their motion. **Dark Matter** exists in an exotic form completely separate from gaseous clouds, planets, and stars, but we know its presence is vital to explaining how fast galaxies move and spin, and our Milky Way Galaxy is no exception. Using various techniques to measure the mass of a galaxy, Astronomers find that visible matter in a galaxy alone can not explain the great motions being demonstrated. Scientists describe **dark matter** as a halo surrounding the galaxy and pulling the matter along.

Dark Matter also has a major presence in groups of galaxies arranged in clusters. While the techniques to determine mass in intergalactic clusters is more complicated, scientists can still confirm that dark matter has to be responsible for a cluster's motion. Dark matter influences the cluster by surrounding all the galaxies, while also still driving their motion individually.

The behavior presents a challenge since even using the process of elimination to determine what it is doesn't help. This matter doesn't is what's called **baryonic**, meaning it's not made up of protons or neutrons, since scientists would be able to detect it then. It is also not **antimatter**, matter made up of antiparticles with equal mass to their normal particle counterpart, but opposite in charge and other quantum properties. Antiparticles would completely annihilate normal particles in galaxies, while even in low quantities would give off a gamma ray signature from this interaction. However we don't see such interaction.

Even black holes, which are indeed dark, can't explain the galactic motion observed, and there are techniques to measure the strength or presence of a black hole through gravitational lensing, or the bending of the spacetime around it. It is truly an unexplainable substance.

So I bet what you want to ask is, "Well then what could it be?" My Senior project in Highschool involved researching potential candidates, though all of these haven't moved past candidate status for the past 15 years. Here are the top 2:

Dark Matter Candidates:

WIMPS

You got to love how physicists come up with these names! This amusing acronym may make you think I'm poking fun or making a joke, but I assure you this is real. The **Weakly Interacting Massive Particles (WIMPs)** may be a leading candidate for **dark matter** due to their incredibly 'weak' interaction with normal, baryonic matter that we observe. These particles would also have a mass 10 to 100 times that of a proton (hence the "massive" component). This combination could be the missing link in explaining the extra mass needed in galaxies to achieve their motion. However, evidence of this particle has yet to be discovered.

Neutrinos

While sounding like a snack you pack in a lunch box, Neutrinos are subatomic particles that are very small, around the size of an electron, and have no charge. They also do not interact with normal matter. Different neutrino variants are known about already, but a theoretical new type of neutrino is thought to exist, that instead of not interacting with normal matter, would interact through gravity. This particle also remains theoretical.

Many more candidates exist out there too, with properties that are only limited by our imagination. What's exciting about this mystery is that dark matter could be anything. Who says **dark matter** has to be subatomic, it could be a particle the size of a basketball, evenly distributed throughout galaxies. These phenomena truly allow us to expand our imagination.

9

The Universe is most likely "Flat"

As we move on to our last revelation, let's first be reminded that we've seen a brief glimpse of how the Universe may have behaved around its birth, but what if we could see an equally fascinating glimpse into the Universe's end, if it even has an end? Not only that, but what I told you that the shape of the Universe could determine its fate?

In this section, we attempt to see into the ultimate fate, and to make the claim that the Universe has a shape, and this shape most likely is mathematically **'flat.'**

So what does that even mean? Well, since we've now have solid evidence that our Universe is accelerating, the big question is, what is the Universe expanding into? I definitely wondered this too, but later I discovered that while the Universe itself is indeed expanding, it may not actually be expanding *into* anything.

So I know I may have lost some of you there, but to assure you that I'm not speaking in riddles, let me take a step back. Common thinking by the layperson would assume that the Universe is like a sheet of paper, like one you may have lying flat on your desk, which as we know has 4 sides and occupies a finite space. In this analogy, the Universe, our sheet of paper, is expanding in all directions, and the edges of our paper are growing in size into uncharted territory. In this Universe, you could draw 2 parallel lines and they would never cross. But as we will soon see, there's more to the story which will provide us with an even better picture.

There is also the common thinking that the Big Bang was an explosion at 1 point and rushing out from that point, however as scientists notice and measure the expansion rate and velocity of galaxies, they find that there is no 'origin' from which these galaxies are accelerating. Instead, the galaxies are expanding away from each other in all directions. Moving away from our paper example, since our Universe is not 2 dimensional, another popular analogy I've heard was raisin bread when it's baking. The whole loaf is expanding along with the separations of the raisins in the bread.

If the Universe is expanding into something, there must be an "edge" associated with the Universe as with our paper. Surprisingly this actually doesn't have to be, and most likely isn't the case with our Universe, as there is no origin. Instead, the Universe can have what's called a **flat euclidean** shape, rather than a spherical or hyperbolic shape. Returning to our sheet of paper example, imagine that we take 2 opposite sides of the paper and attach them together, edge to edge, so that their edges meet. We've now formed a cylinder. If an ant were to travel along the width of the cylinder's surface, it would never meet an edge we just attached together.

Now the fun part, and please stay with me. If we connect the ends of the cylinder together, we now form a donut-shaped object, or in mathematical terms, a **torus**. This same ant can travel anywhere on the surface of our donut, but never see an edge. All of the Universe is on the Donut, and while the donut could grow bigger, the universe itself is contained with no edge.

You may, or may not, have understand the above paragraph. But even if you do, you have another question: "Why would the Universe be considered 'flat' if it exists in this multi-dimensional curved shape?" The reason I used the term 'mathematically flat' is due the torus shape still being **"Euclidean"**, or the shape of the Universe behaves in an Euclidean manner, such as the angels of a triangle sum to 180 degrees, and properties of parallelograms being intact. I brought up drawing parallel lines because in **Euclidean space**, 2 parallel lines would never intersect, even in our donut Universe. Hope that now makes sense!

It's indeed difficult for us to envision a multi-dimensional Euclidean geometry in which our Universe resides, but you now see how the Universe is always connected to itself, and how you could theoretically travel anywhere. In this shape, there is no edge of the Universe where you can sit and observe the unknown that is uncovered once the Universe expands a little more.

If you went far enough, depending on the shape, you may even return to where you were. If you know the diameter of the Universe, shine a beam of light across it and follow it, you would see a version of your galaxy from billions of years ago, which you'd be able to calculate. What's also mind-blowing is you could continue to follow the beam of light to make a second round trip, wrap around again, and see an earlier version, and so on and so forth.

Evidence for our Universe's flatness comes mainly from the Universe's **critical density**, the average value of baryonic matter in a cubic meter, in the Universe. Understanding this density is extremely vital to our determination of our Universe's shape. If our measured density equals that of the theoretical critical density, we are in a flat university that will eventually stop expanding, over infinite time. If our density is greater the the critical density, then there is enough gravity to eventually stop the expansion and actually pull the Universe in and collapsing itself, know in science groups as the **Big Crunch,** which would suggest a closed Universe. Majority of measurement conducted so far point to a Universe that has a density slightly below the critical density, which suggests that the Universe is indeed flat and will for expand for eternity. In our donut analogy, the donut is ever increasing in size.

However emerging results are being discovered that indicate that we may actually be in a closed Universe. Observations recorded by scientists working with the Planck Space Telescope, suggests that the significant gravitational lensing curvature noticed in their data could only be explained by a closed Universe. There is also talk that a closed Universe could make more sense when the presence of dark matter and dark energy is accounted for in regions of space. As the debate remains unsettled on if the Universe is closed or flat, this will no doubt be the focal point of many astronomical surveys and discoveries in the future.

10

Conclusion

With the mind blowing content finally coming to a close, I hope you've enjoyed this tour through our Universe! It brought me great joy in writing about these topics, fleshing out concepts that took me years to get, in a format you could enjoy. Many of these cosmic details still get to me, especially as I look up to the heavens at night. Our Universe is indeed full of countless more stories.

I also hope this inspired you to wonder on your own about this world, and possibly create your own list of awesome and mind-bending facts. Lastly, I hope that this book completed its goal of supplying you with at least one new fact that you never knew, and one that would impress your friends and family!

If you found this book helpful, please leave a favorable review for this book on Amazon! Thank you again!

11

Resources

Are we made of stardust? (2018, June 4). Natural History Museum. https://www.nhm.ac.uk/discover/are-we-really-made-of-stardust.html#:~:text=Most%20of%20the%20elements%20of,originated%20from%20the%20Big%20Bang.

Ask an astronomer. (n.d.). Cool Cosmos. https://coolcosmos.ipac.caltech.edu/ask/8-How-far-away-is-the-Sun-#:~:text=The%20Sun%20is%20at%20an,8%20minutes%20to%20reach%20us.

Cain, F. (2015, March 6). How far back are we looking in time? *phys.org.* Retrieved October 14, 2023, from https://phys.org/news/2015-03-how-far-back-are-we.html#:~:text=If%20we%20had%20a%20telescope,than%204.2%20light%2Dyears%20away.

Cosmic Speed Limit | AMNH. (n.d.). American Museum of Natural History. https://www.amnh.org/exhibitions/einstein/light/cosmic-speed-limit#:~:text=But%20Einstein%20showed%20that%20the,can%20travel%20at%20that%20speed.

Dark Energy and Dark Matter | Center for Astrophysics. (n.d.). https://p web.cfa.harvard.edu/research/topic/dark-energy-and-dark-matter# :~:text=Dark%20matter%20pulls%20galaxies%20together,of%20dark %20energy%20in%201998.

Dark energy, Dark matter - NASA Science. (n.d.). https://science.nasa.go v/astrophysics/focus-areas/what-is-dark-energy/

Forces | Universe – NASA Universe Exploration. (n.d.). NASA Universe Exploration. https://Universe.nasa.gov/Universe/forces/#:~:text=Th ey%20understand%20that%20there%20are,shaping%20the%20Unive rse%20we%20inhabit.

Most distant galaxy from Earth could be giving birth to earliest stars. (2022, April 7). https://www.nhm.ac.uk/discover/news/2022/april/scientis ts-find-the-most-distant-object-ever-seen-from-Earth.html#:~:text= The%20most%20distant%20object%20ever%20seen%20from%20Eart h%20may%20have,beyond%20the%20current%20record%20holder.

Muro, T. (2022, December 26). Cosmic timeline: What's happened since the Big Bang. *Science News Explores.* https://www.snexplores.o rg/article/what-happened-since-big-bang-physics-Universe-cosmic- timeline#:~:text=10%2D32%20to%2010%2D10,weak%20nuclear%20 and%20electromagnetic%20forces.

Notification. (n.d.). https://www.discovery.com/science/Universe-Is- Expanding

Rayne, E. (2022, November 13). What is the shape of the Universe? *livescience.com.* https://www.livescience.com/what-is-shape-of-Univ erse#:~:text=So%2C%20what%20is%20it%20shaped,angles%2C%20e xperts%20told%20Live%20Science.

Riess, A. (2009, February 18). *Dark energy | Definition, Discoverers, &* *Facts*. Encyclopedia Britannica. https://www.britannica.com/science/ dark-energy

Siegel, E. (2022). Ask Ethan: What does "Grand Unified Theory" mean? *Big Think*. https://bigthink.com/starts-with-a-bang/grand-unified- theory/

Sutter, P. (2023, March 26). The Universe might be shaped like a doughnut, not like a pancake, new research suggests. *Space.com*. https://www.space.com/Universe-might-be-shaped-like-doughnut- not-pancake

Svs. (2014). How comets are born. *SVS*. https://svs.gsfc.nasa.gov/ 11693#:~:text=Astronomers%20believe%20comets%20materialized% 20more,water%20ice%20began%20clumping%20together.

The Big Bang. (n.d.). Hubble. https://hubblesite.org/contents/articles/ the-big-bang#:~:text=About%2013.8%20billion%20years%20ago,Ato ms%20formed%2C%20then%20molecules.

Tillman, N. T., & Pultarova, T. (2023). What is dark matter? *Space.com*. https://www.space.com/20930-dark-matter.html

What shape is the Universe? A new study suggests We've got it all wrong. (2020, September 4).
   Quanta Magazine. https://www.quantamagazine.org/what-shape- is-the-Universe-closed-or-flat-20191104/